Airplanes the Second World War

Coloring Book

War Planes Coloring Book for Kids

War Planes 2

Forester Wood

ISBN: 9798644053865

RF U
P3975
WINGS FOR VICTORY

KEEP'EM FIRING

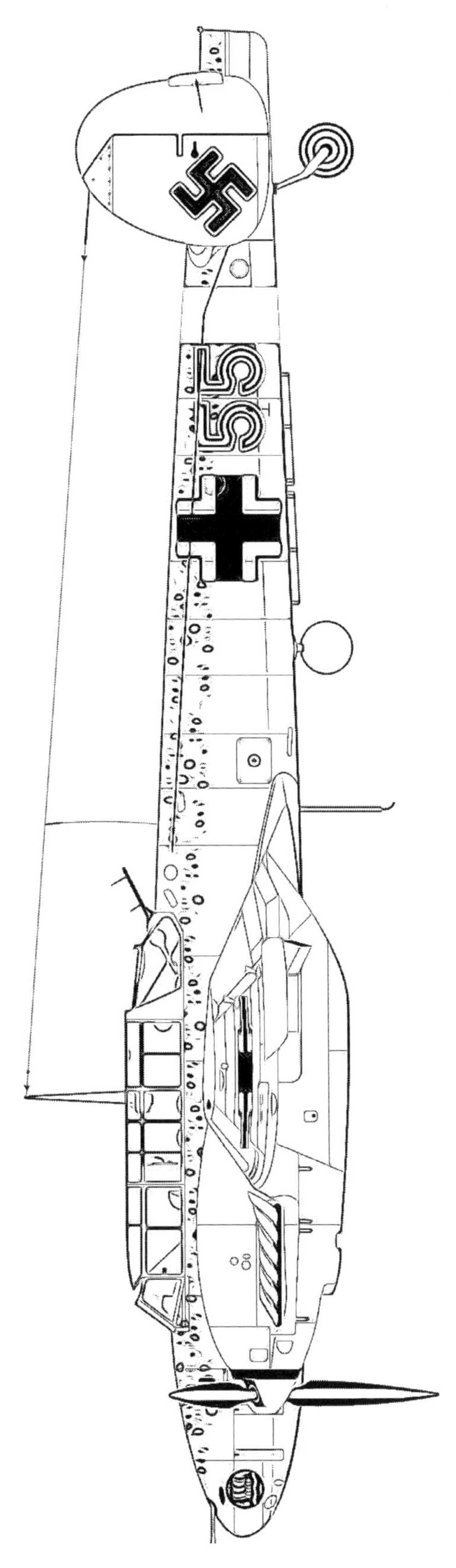

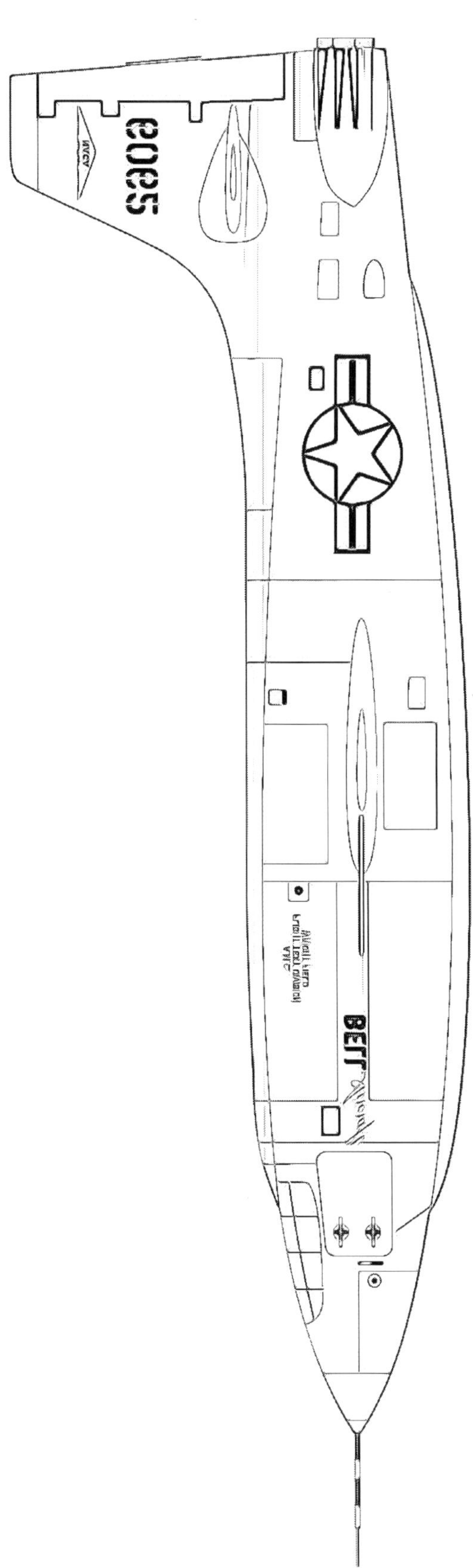

RF
U
P3975

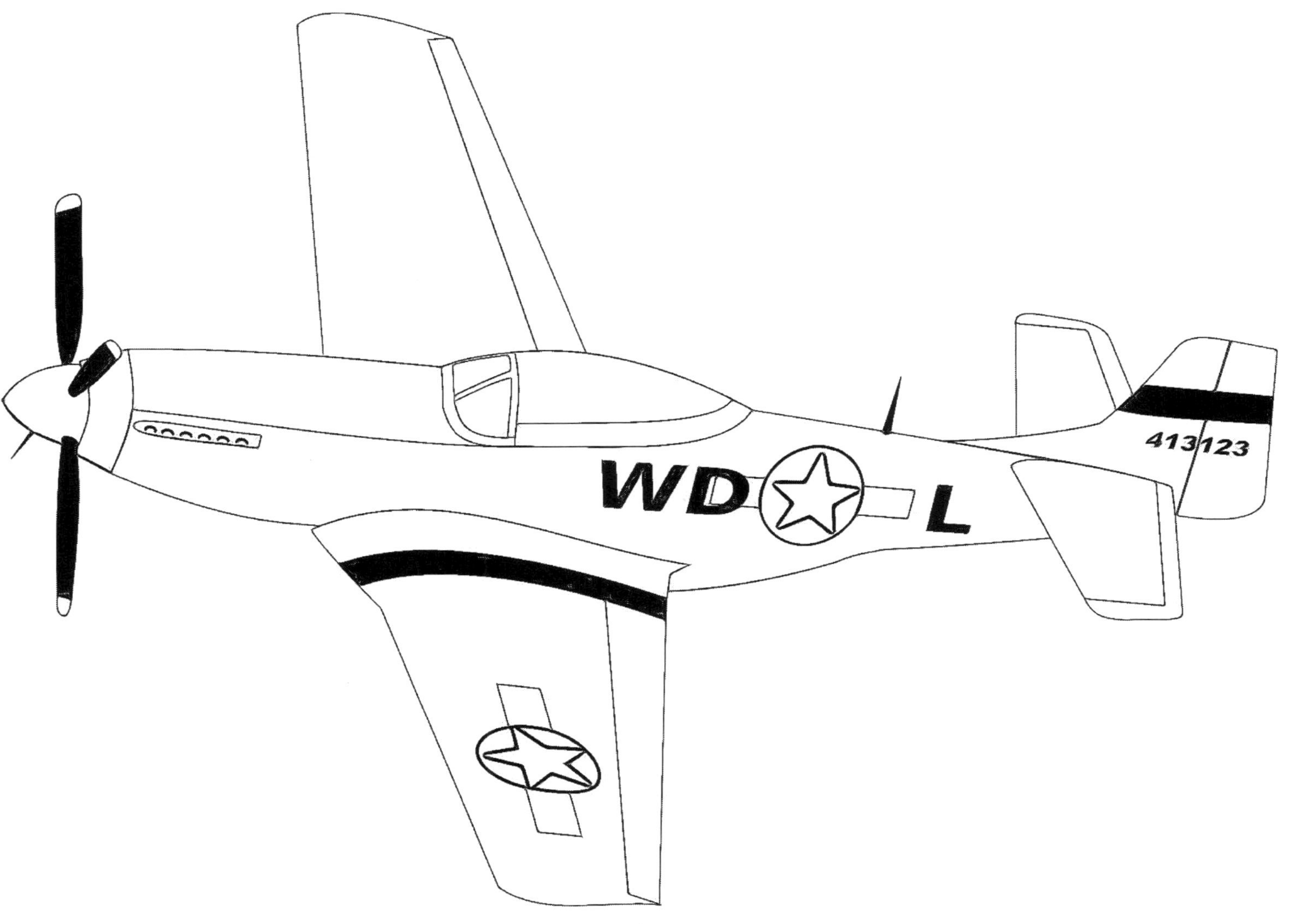
WD
L
413123

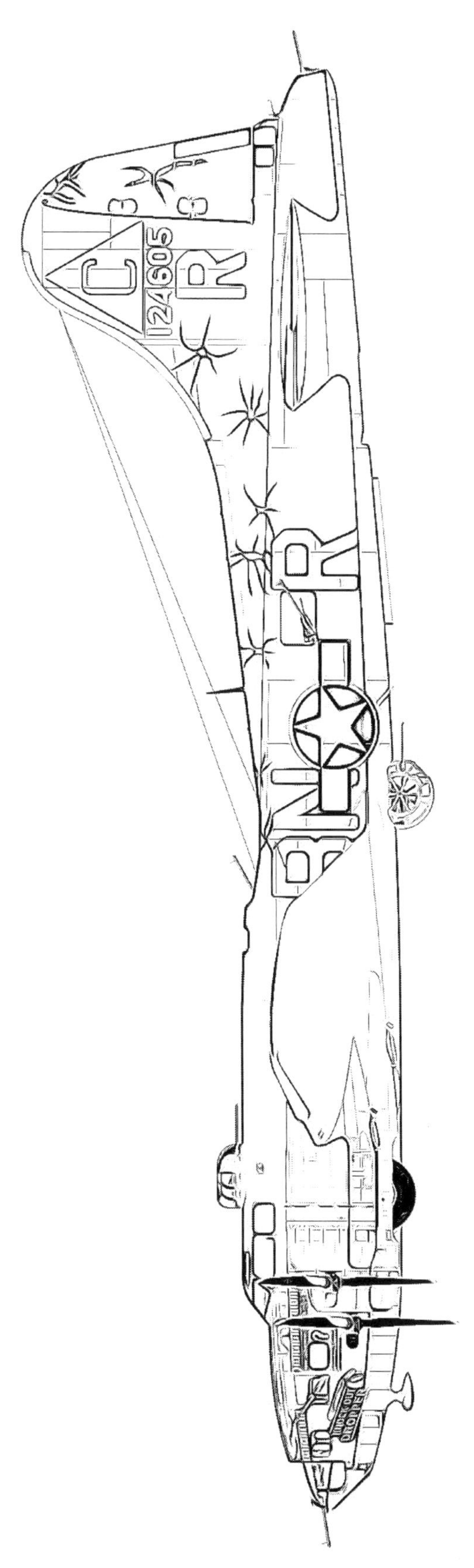

124605

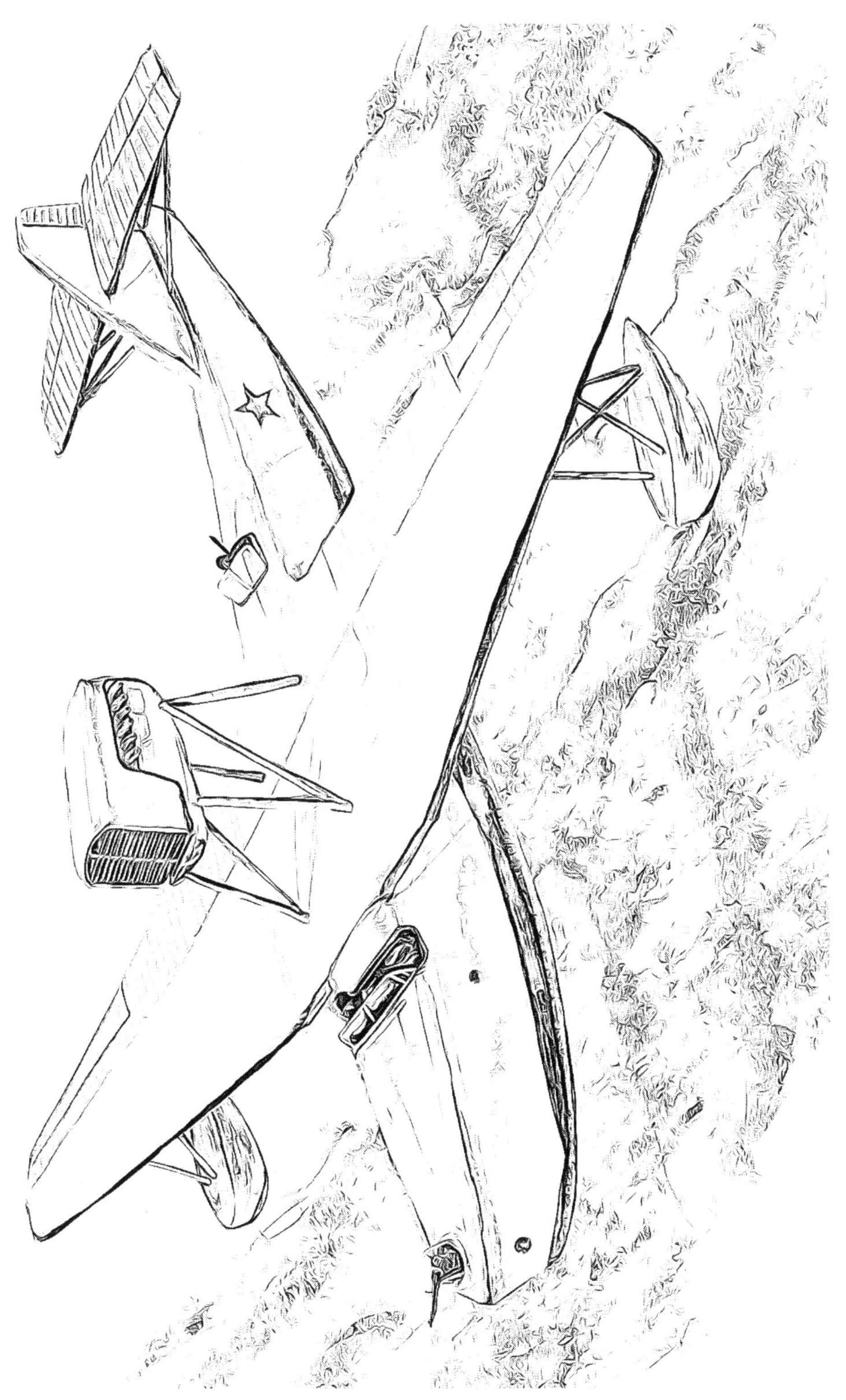

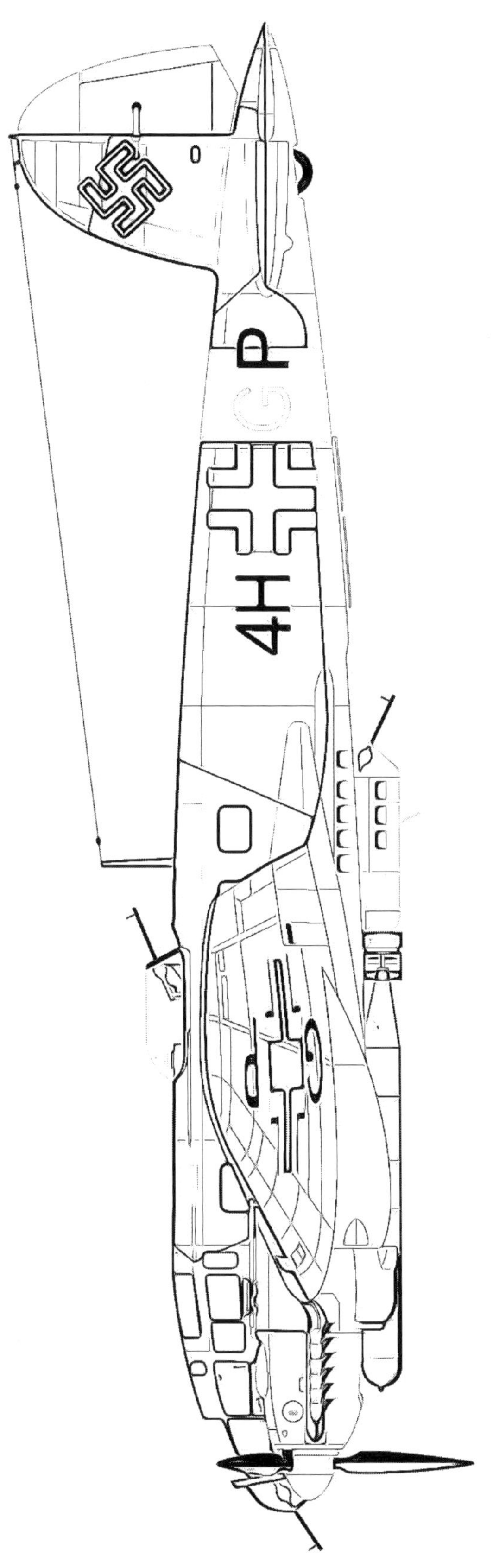

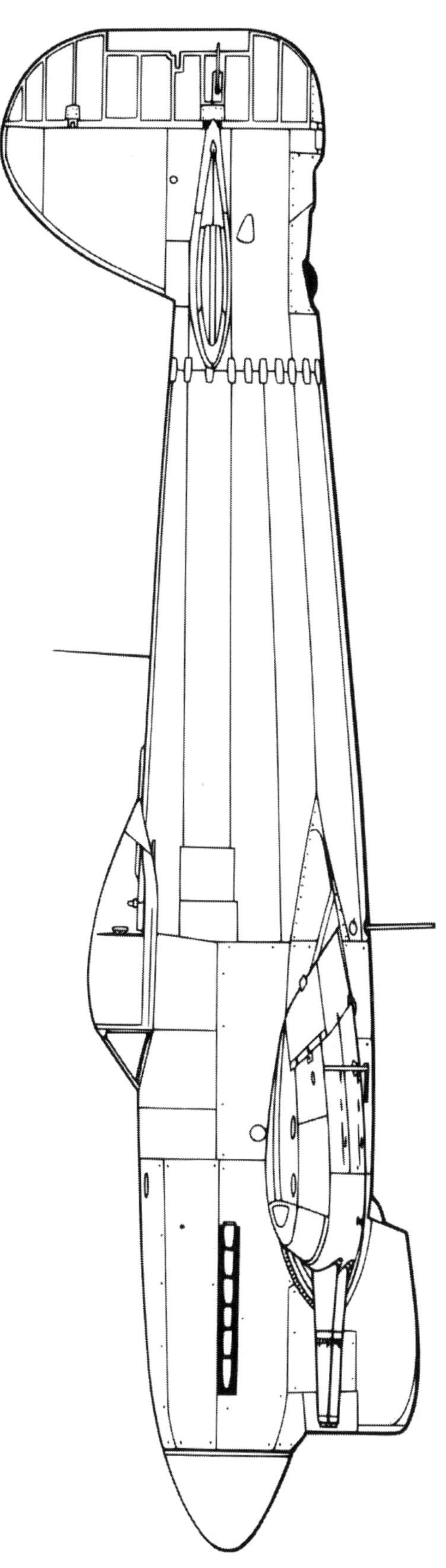

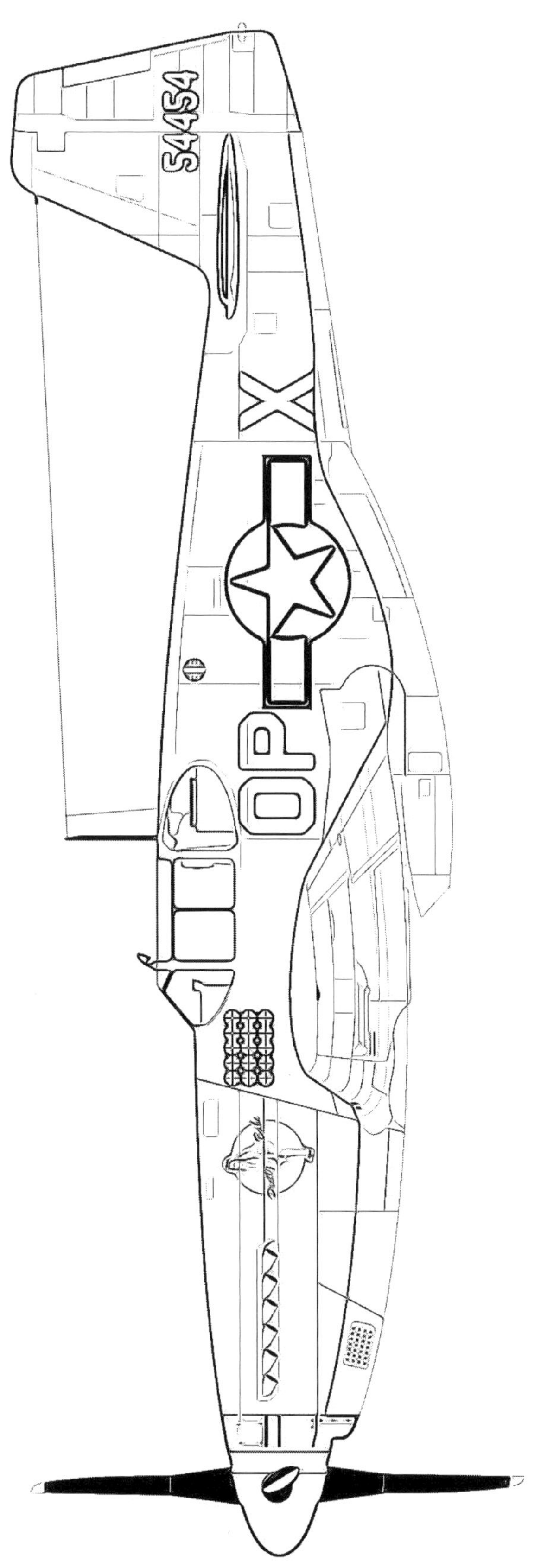
54454
X
OP

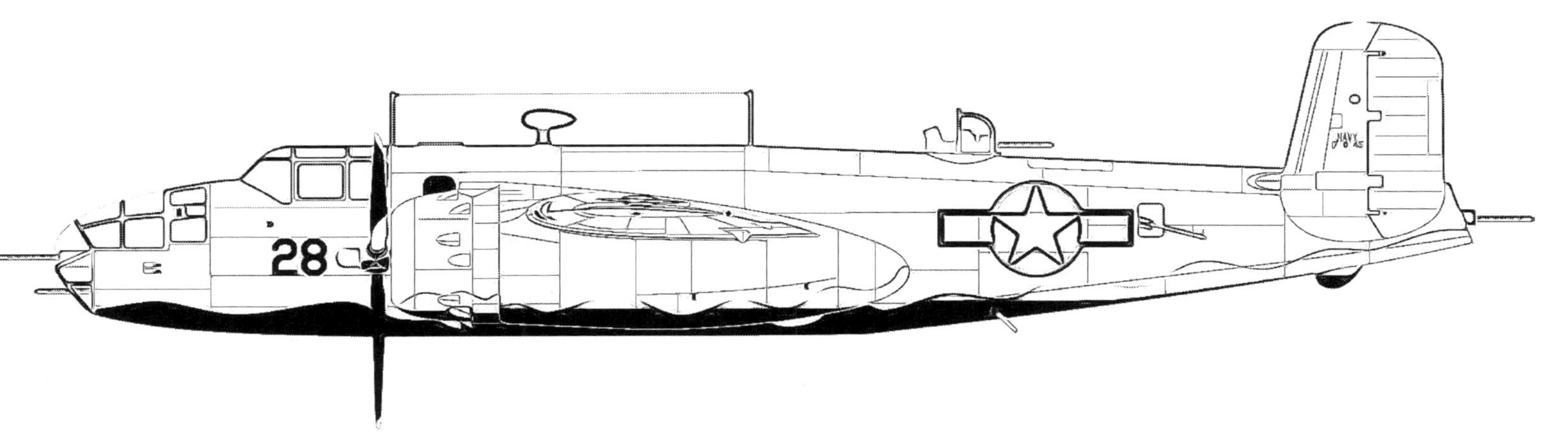
28

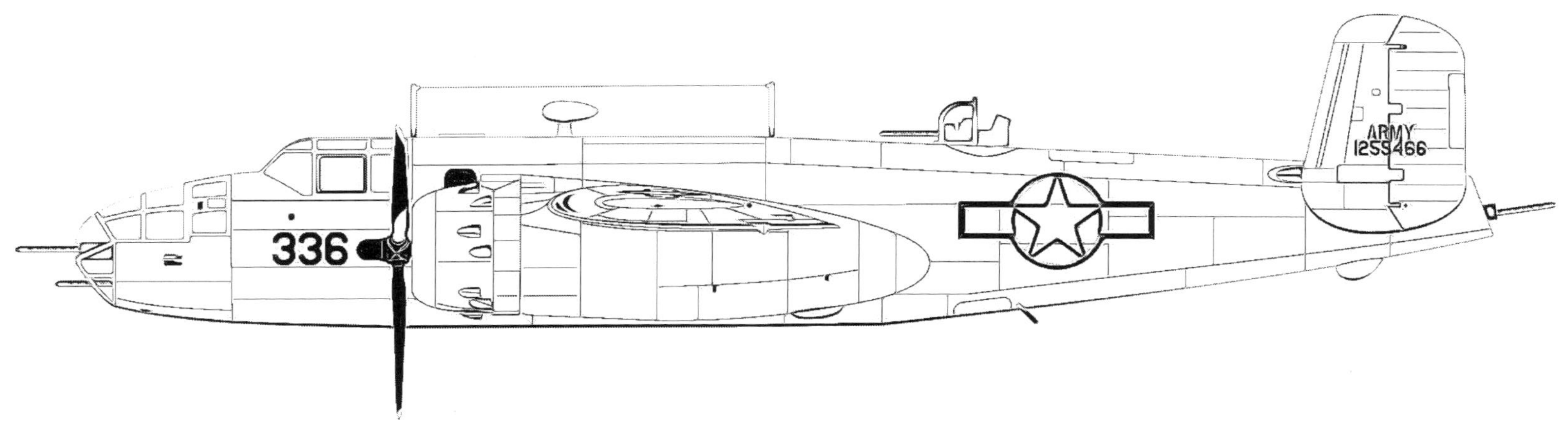
ARMY
336

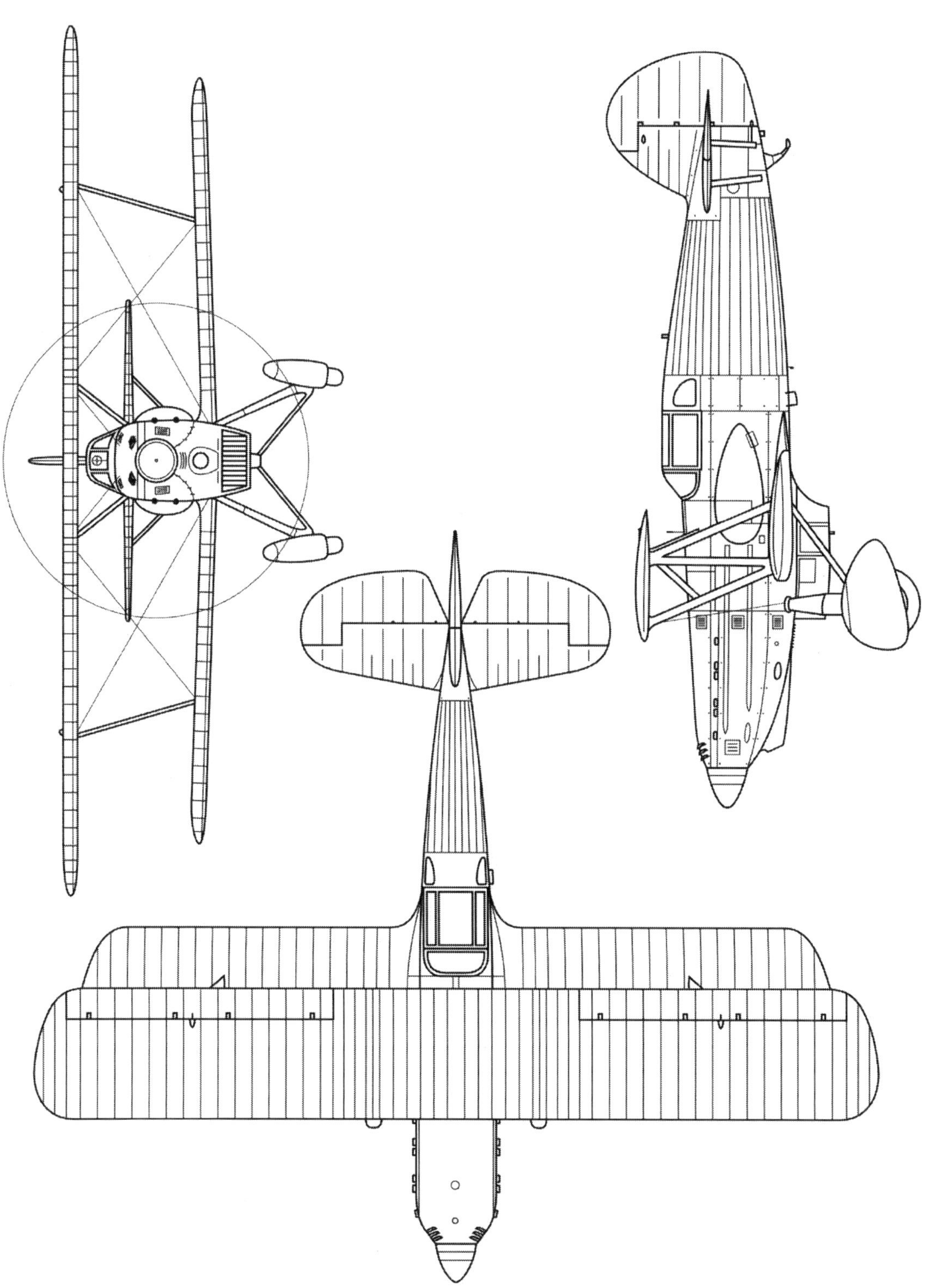

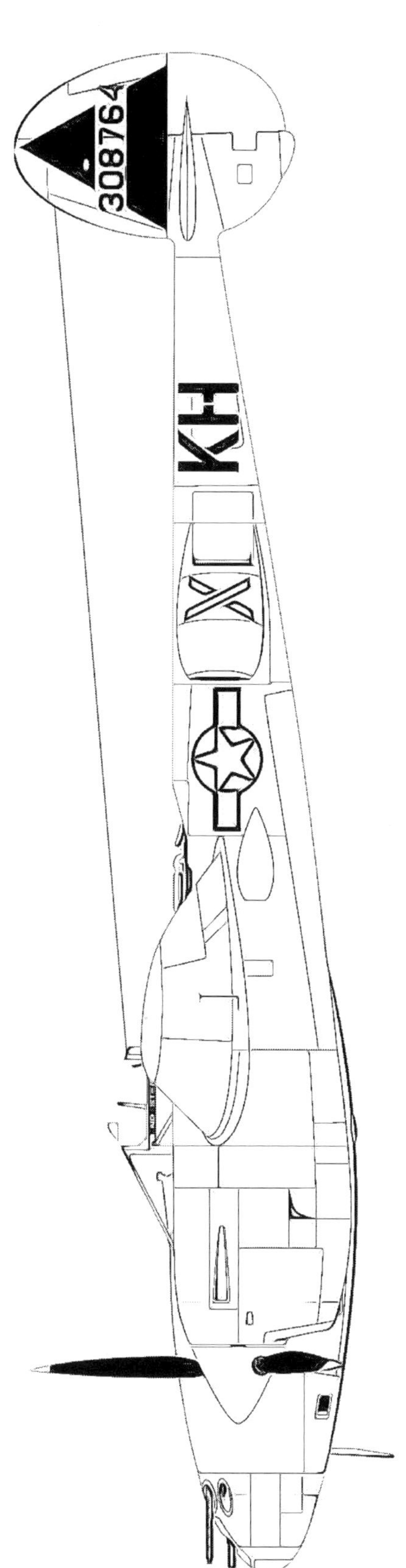

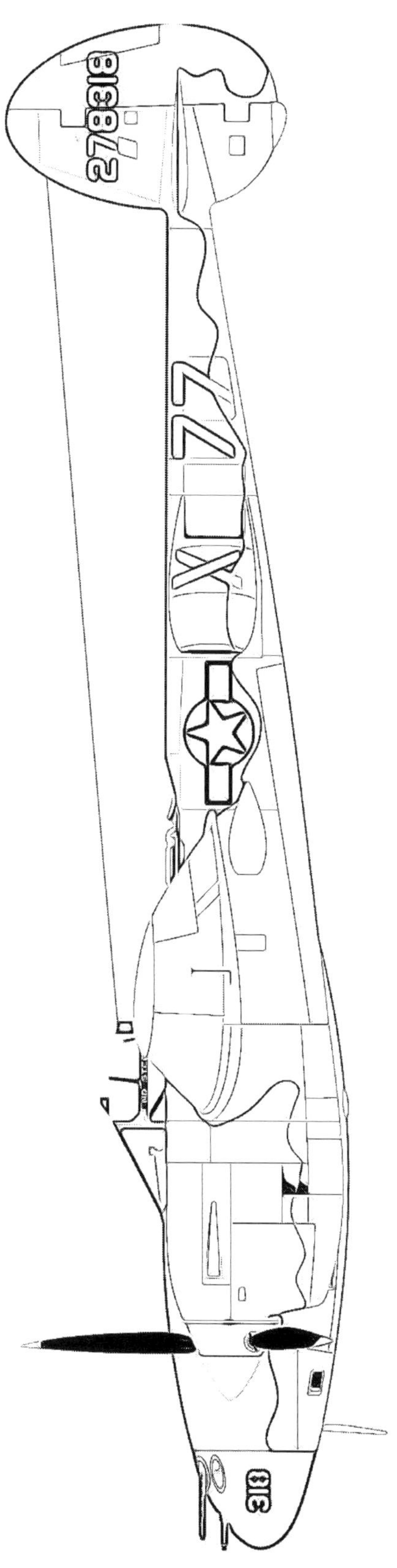

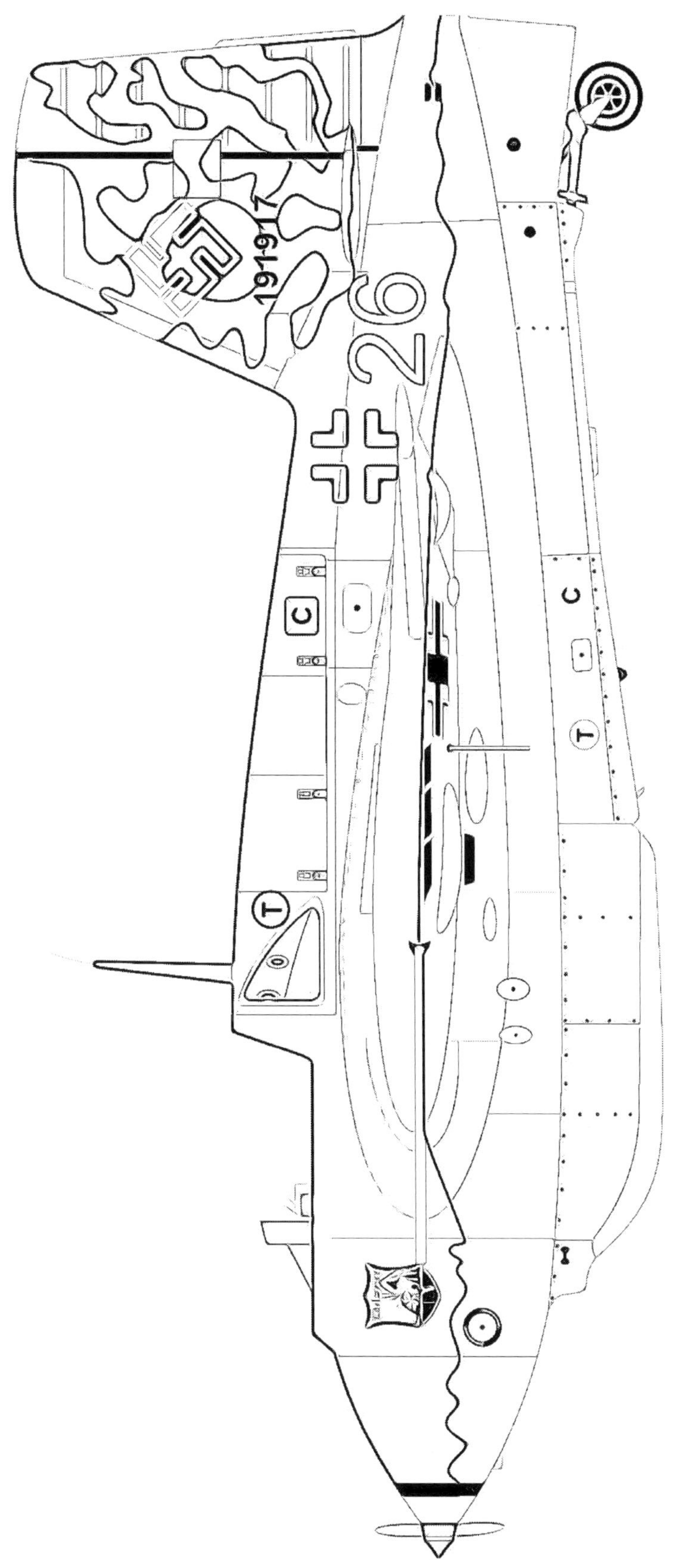

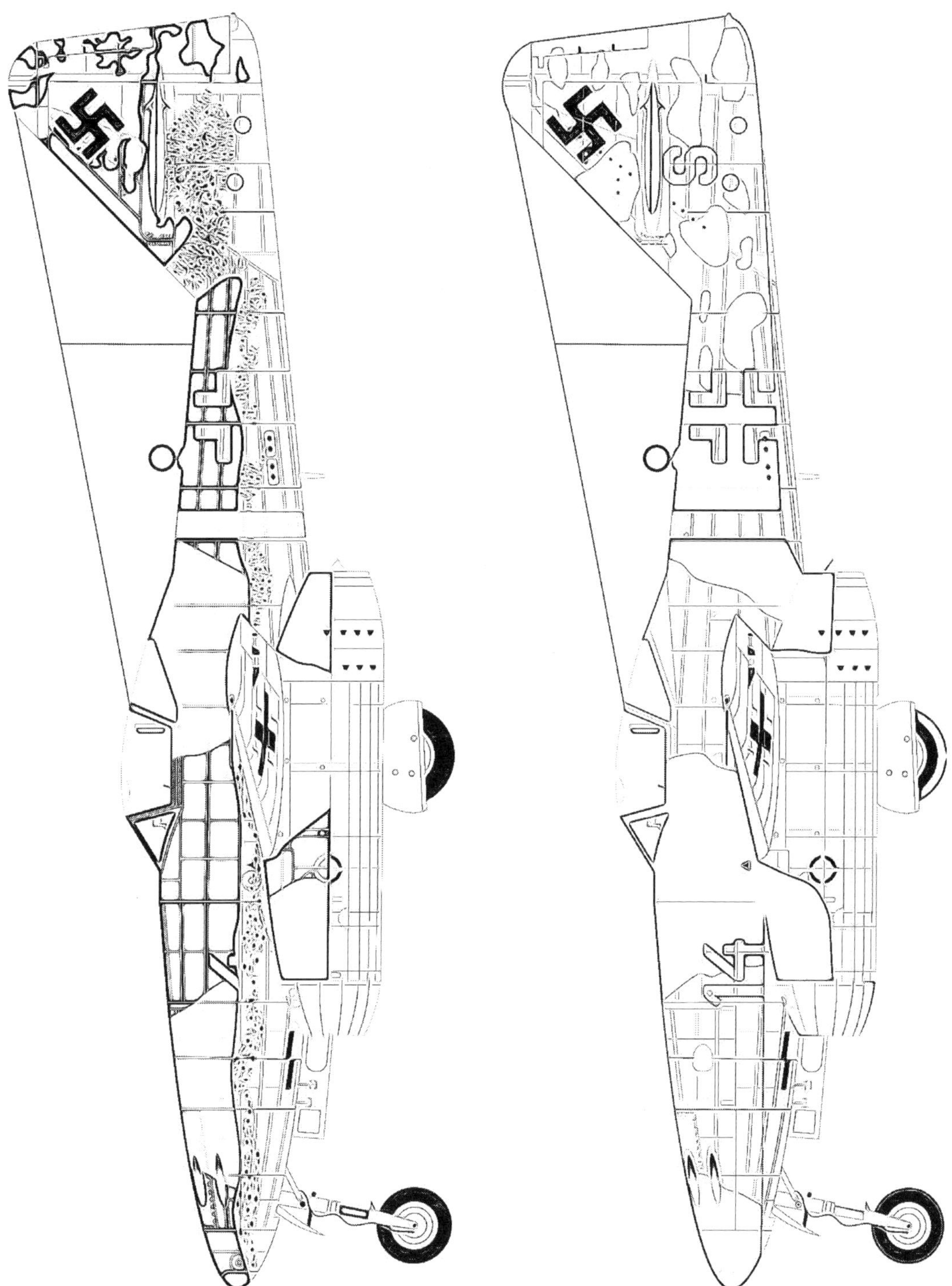

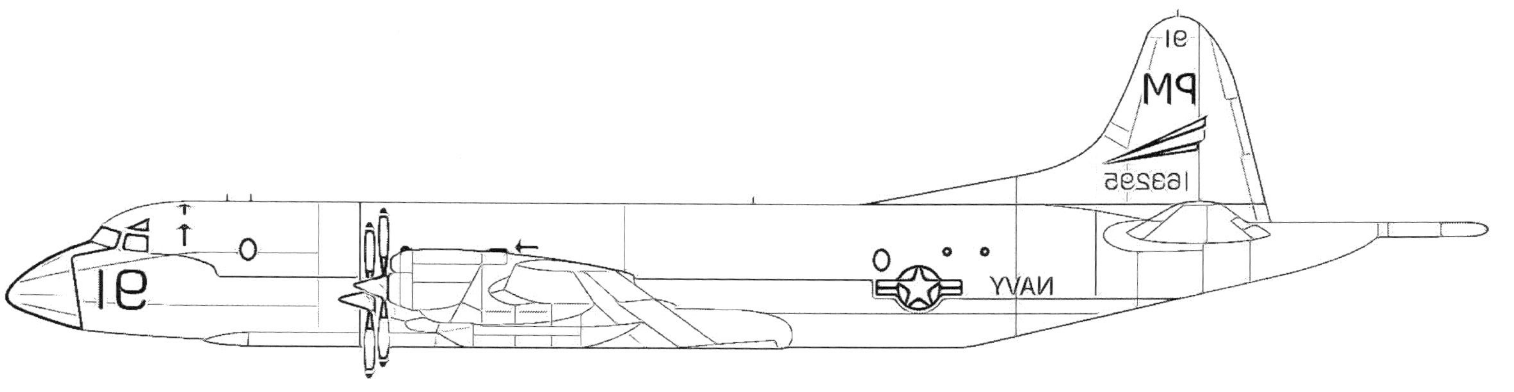
PM
NAVY

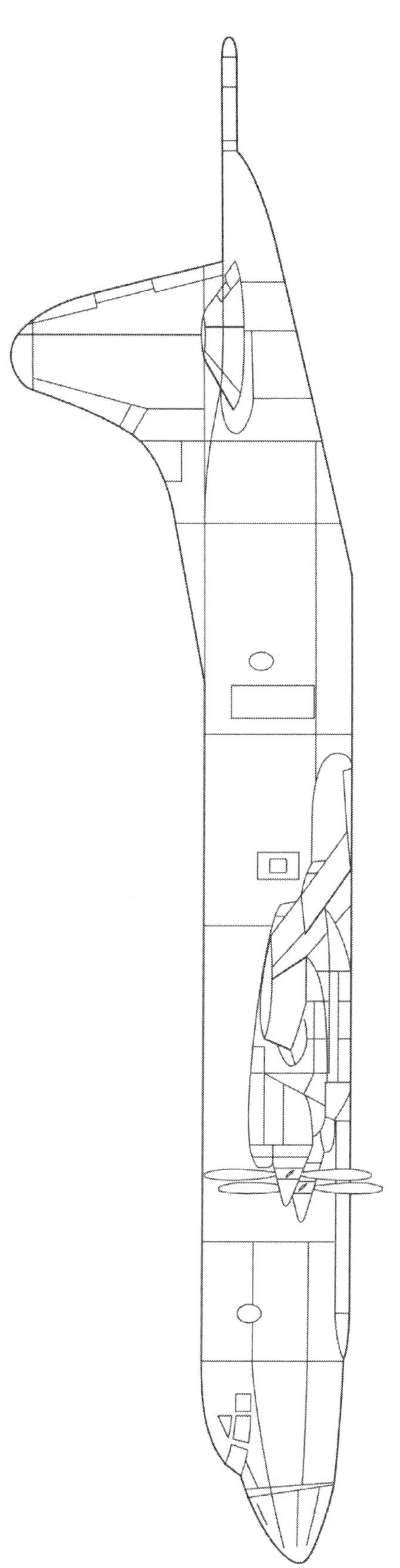

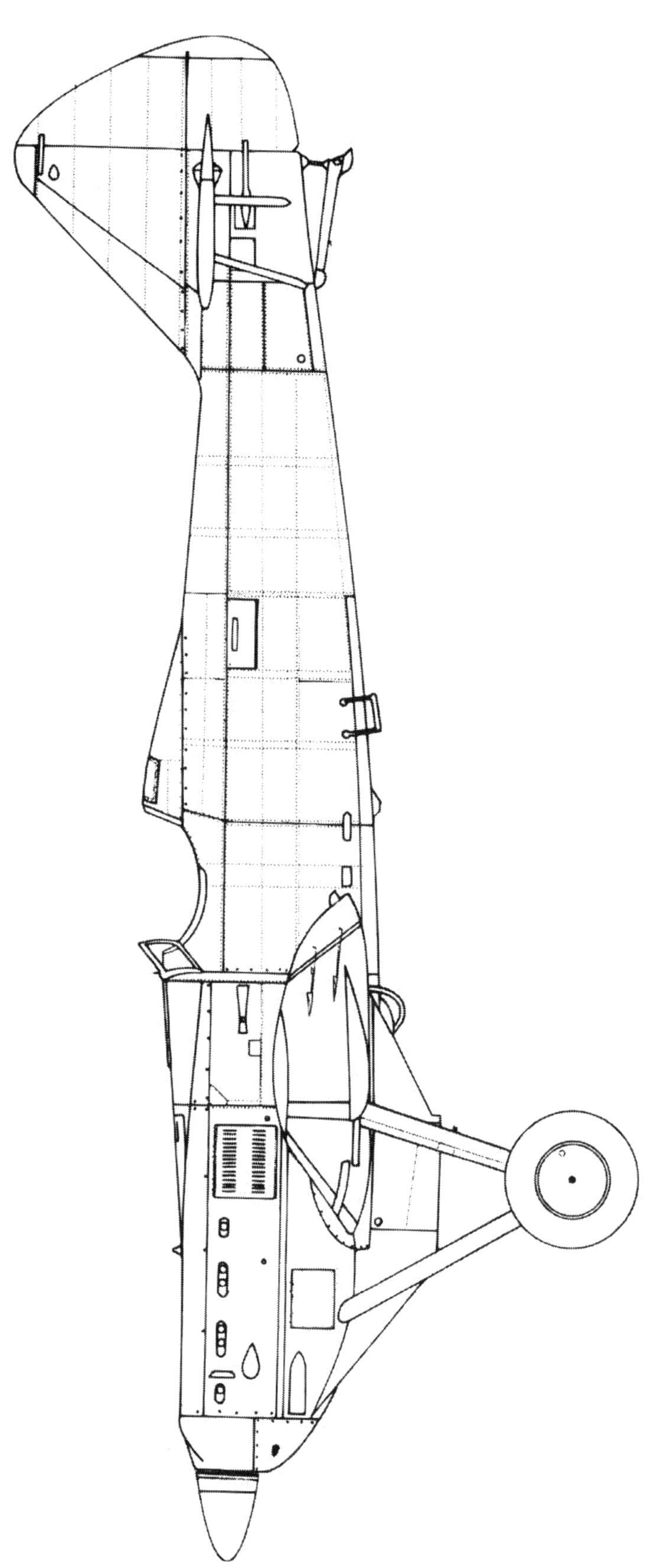

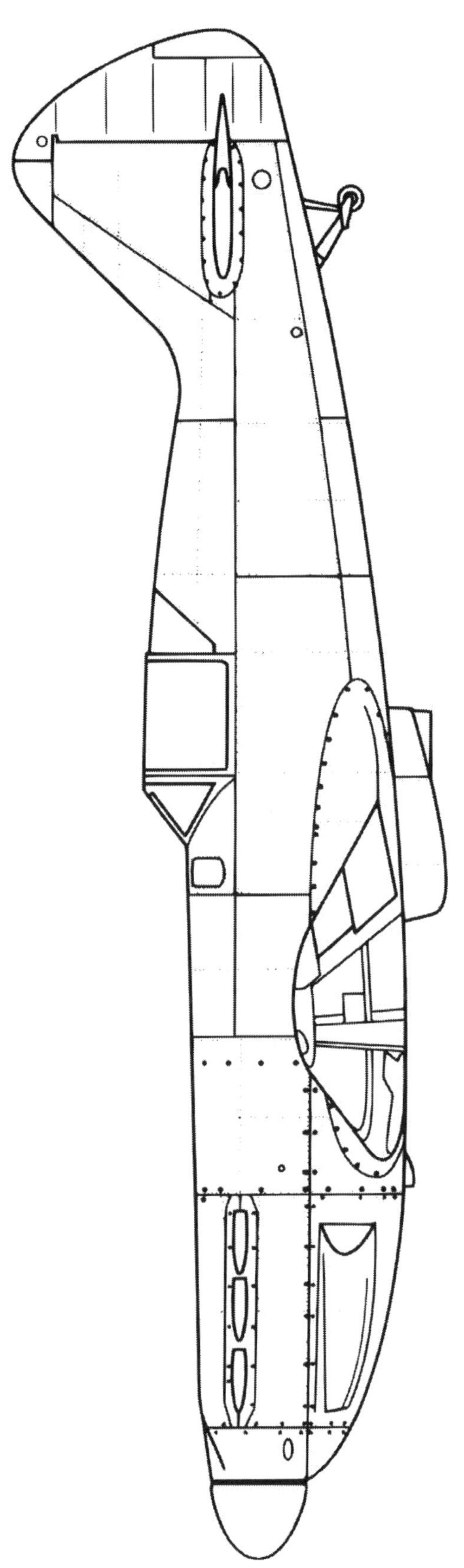

VR A
KB726

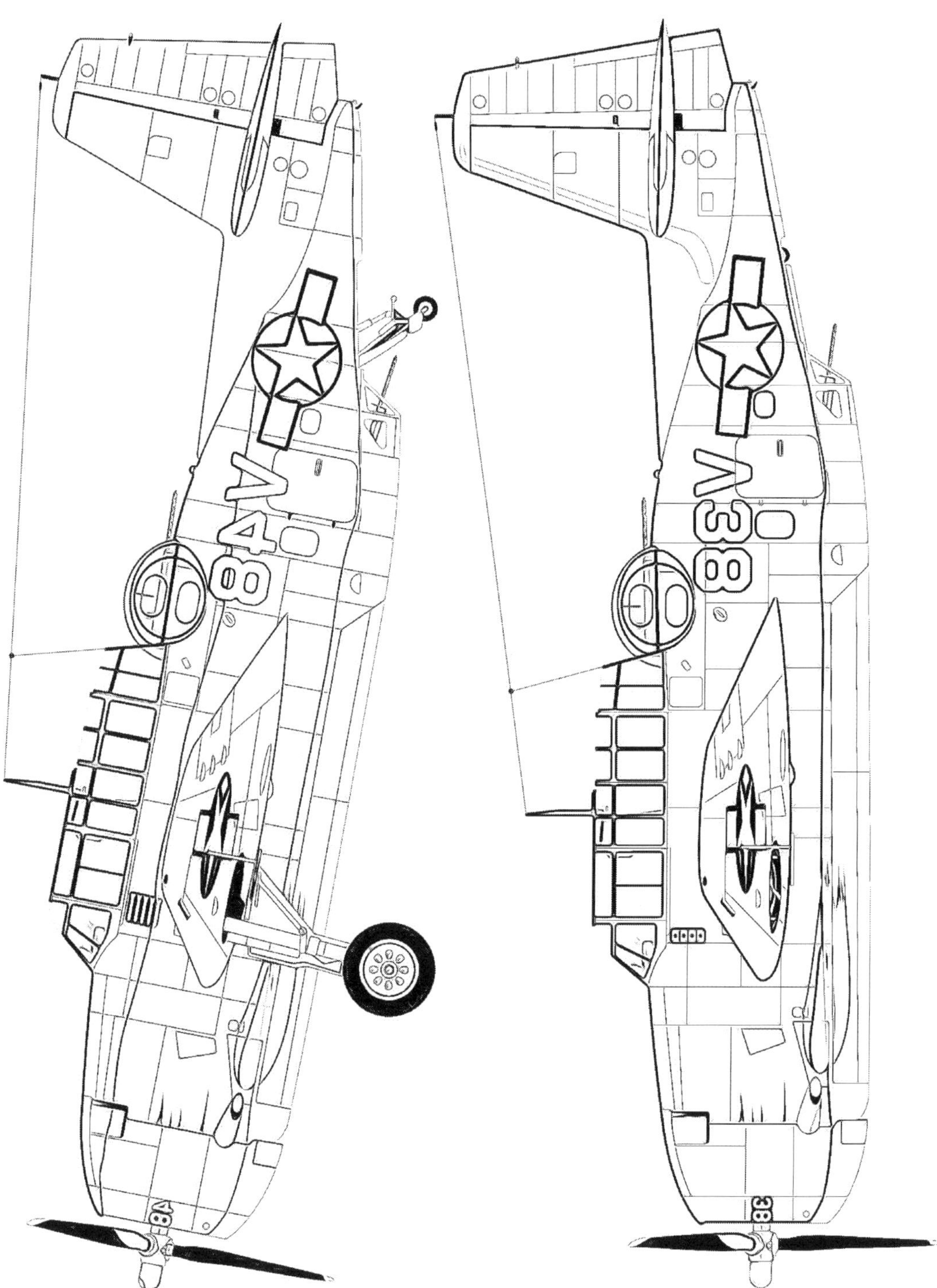

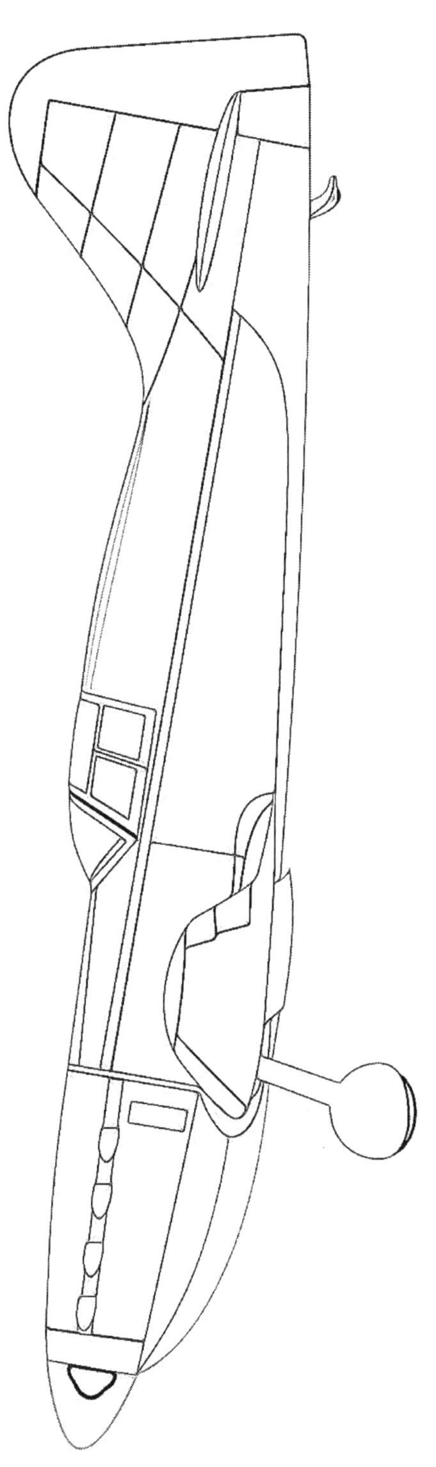

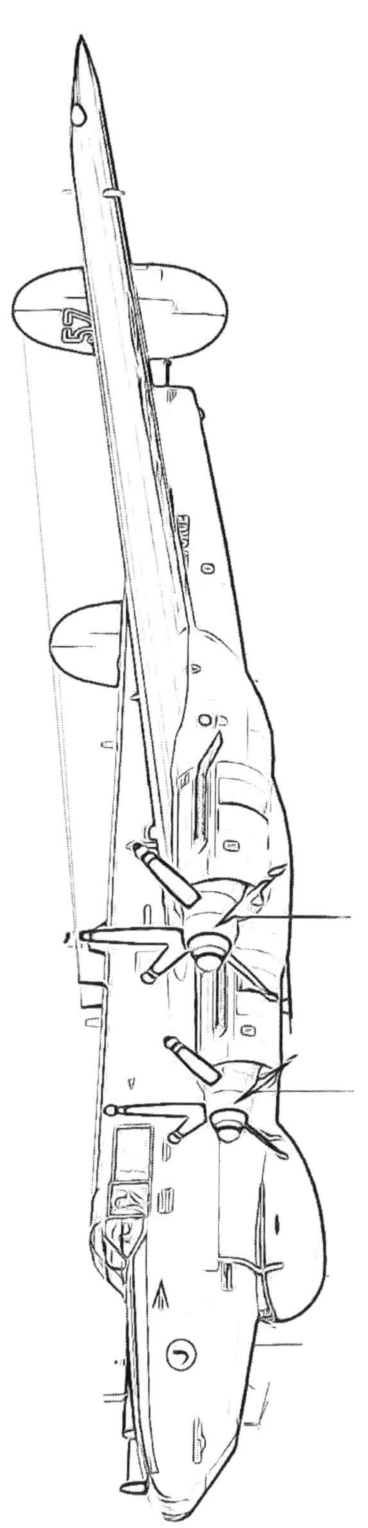

Printed in Great Britain
by Amazon

51414934R00036